CONTENTS

A WORD FROM THE AUTHOR

Welcome to Pymm, a land inhabited by humans, elves, half-elves, dwarves, and even Minotaurs. Here, kings rule, the Knights of the Golden Sword are the bravest heroes of the land, and armies small and large fight off invaders. Pymm is also a land where dragons and other strange creatures (such as the repulsive, monstrous glubs) are familiar menaces, wizards use their great powers for good or evil, and common people (stone workers, millers, bridge builders, cart drivers, ferry operators, and so forth) go about their daily lives finding amusement at fairs and in the festivities at the castles of the local kings.

As you wander through these eighty-plus puzzles set in Pymm, you will find that some are easy and some that are genuinely difficult. Hopefully, you will find many that irresistibly challenge you. My personal favorites are marked with asterisks.

The puzzles are not arranged in any order of difficulty. This has been done deliberately so that all puzzle solvers, from novices to experts, will be interested in perusing the entire book. There are hints to help you solve some of the puzzles, but use these clues only if you must. First give yourself a chance to solve them on your own. Many of the puzzles can be solved without using algebra, and some require only the most basic algebra knowledge.

Suggestions and responses are not only welcome, they are wanted. It's especially nice when a reader finds an easier or more interesting solution to a puzzle.

—Margaret C. Edmiston

FANTASTIC BOOK OF Math Puzzles

Margaret C. Edmiston

Illustrated by
Jim Sharpe

Sterling Publishing Co., Inc.
New York

For my mother, the wisest woman I have ever known.

ACKNOWLEDGMENTS

With great thanks to Martin Gardner, J.A.H. Hunter, Boris A. Kordemsky, Geoffrey Mott-Smith, and Robert Muller for their inspirational writings, and to my friends Garland Turner and Kathryn Wallo and my brother, Steve Edmiston, for their help and encouragement.

Special thanks to the illustrator, Jim Sharpe.

Library of Congress Cataloging-in-Publication Data

Edmiston, Margaret C.
Fantastic book of math puzzles / by Margaret C. Edmiston ; illustrated by Jim Sharpe.
p. cm.
Includes index.
ISBN 0-8069-6114-7
1. Mathematical recreations. I. Title.
QA95.E355 1997
793.7′4—DC21 97-15300
CIP

10 9 8 7 6 5 4 3 2 1

First paperback edition published in 1998 by
Sterling Publishing Company, Inc.
387 Park Avenue South, New York, N.Y. 10016

Distributed in Canada by Sterling Publishing
% Canadian Manda Group, One Atlantic Avenue, Suite 105
Toronto, Ontario, Canada M6K 3E7
Distributed in Great Britain and Europe by Cassell PLC
Wellington House, 125 Strand, London WC2R 0BB, England
Distributed in Australia by Capricorn Link (Australia) Pty Ltd.
P.O. Box 6651, Baulkham Hills, Business Centre, NSW 2153, Australia
Manufactured in the United States of America

Sterling ISBN 0-8069-6114-7 Trade
0-8069-8669-7 Paper

Chapter I

GETTING STARTED

SLAYING GLUBS

Pymm is inhabited by an unknown number of hideous monsters known throughout the land as glubs. Glubs live underground but can rapidly burrow to the surface if they smell a human—one of their favorite treats.

Between them, Grabus and Hylar, two Knights of the Golden Sword, have slain twenty-four glubs. Grabus has killed four more glubs than Hylar has killed. How many glubs has each man slain?

Answer on pages 81–82.

HOW MUCH DID ALARANTHUS WEIGH?

Pymm has many dragons. A few years ago, one of these dragons, Alaranthus, though not fully grown, weighed 1000 pounds plus two-thirds of his own weight. How much did Alaranthus weigh?

Answer on page 74.

HOW MUCH DOES THE FULLY GROWN ALARANTHUS WEIGH?

The fully grown Alaranthus has a weight equal to one-fifth of his own weight plus the weight of the dragon Montal, who weighs 800 pounds less than Alaranthus. How much does the mature Alaranthus weigh?

Answer on page 74.

HOW MANY WINGED CATS?

Gareth collects winged cats for his menagerie of strange Pymmian creatures. One day, a friend asked him, "How many winged cats do you have now, Gareth?"

Gareth answered, "I have two-thirds of their number plus two-thirds of a winged cat." How many winged cats does Gareth have?

Answer on page 74.

THE COST OF CIDER

In Pymm, the cost of a liter of cider plus its liter container is 10 stickels. The cider costs 4 stickels more than the liter container. How much does the cider itself cost?

Answer on page 83.

APPLES FOR ALL

While out exploring, a group of boys came upon an apple tree whose fruits were ripe for the picking. One of the boys climbed the tree and picked enough apples for each boy to have three, with none left over. Then along came three more boys, making it impossible to divide the picked apples evenly.

However, after picking one more apple and adding it to the total, every boy had two apples, with none left over. How many apples were divided among how many boys?

Answer on page 66.

HOW MANY HORSES?

If the warrior princess Mistar were to ascertain how many men and how many horses she has under her command by counting both the legs and heads, she would count 45 heads and 120 legs. How many horses are under Mistar's command?

Answer on page 73.

WOODEN SWORDS

Jarius and Kylar had a sparring match with wooden swords. If Jarius had lost the match he would then have won the same number of sparring matches as Kylar. If Jarius had won the match he would then have won twice as many matches as Kylar.

How many matches had each boy won before this match?

Answer on page 94.

HOW MANY LUCS?

Jarblek, Belgar, Polgar, and Garion were each paid the same hourly rate to build a bridge. One day Jarblek worked the full day, Belgar worked half a day, and Polgar worked half as long as Belgar and a third as long as Garion. Together the four earned 40 lucs. How many lucs did each receive?

Answer on pages 73–74.

WEIGHING A POUND OF FLOUR

"I want exactly one pound of flour," said the customer to the miller.

"Sorry," said the miller. "My scales are faulty. One arm is longer than the other."

"Do you have some lead pellets and a one-pound weight?" asked the customer.

The miller provided these things and the customer was able to weigh a pound of flour.

How did she do it?

Answer on page 91.

HOW OLD IS MONGO'S SON?

"How old is your son?" Blat asked Mongo.

Mongo replied, "If my son's age is increased by 6 years, the result is a number which has a positive integral (whole number) square root. If his age is decreased by 6 years, the result is a number which is that square root."

How old is Mongo's son?

Answer on pages 75–76.

RINGS FOR THE PRINCESSES

There are a number of kings in Pymm, each with his own base of power and his own castle. One of these, King Firnal, had a box containing three gold rings. He wanted to divide the rings among his three daughters so that each received a ring, but one ring remained in the box. How could he do this?

Answer on page 81.

TWO RIDERS

A knight on horseback left Belft to ride to Dalch at the same time another knight left Dalch on horseback to ride to Belft along the same road. The first knight traveled 30 miles per hour and the second traveled 28 miles per hour. How far apart were the two riders one hour before they met?

Answer on page 91.

A LAME HORSE

A knight had ridden one-third the total distance of his trip when his horse became lame. He finished the journey on foot, spending twenty times as long walking as he had spent riding. How many times faster was his riding speed than his walking speed?

Answer on page 64.

THE MESSENGER

A messenger capable of running long distances set out to deliver a message so that reinforcements could be brought to help fight a horde of glubs. The messenger had to run for 24 miles. For two-thirds of the distance, he averaged 8 miles per hour. At what rate did he have to run the remainder of the distance in order to average 12 miles per hour for the entire journey?

Answer on page 86.

JOUSTING TOURNAMENT NUMBER

Each of the eleven competitors in a jousting tournament was given a number between 1 and 11. Sir Bale's son asked his father, "What is your number in the tournament, father?"

Sir Bale replied, "If the number of numbers less than mine is multiplied by the number of numbers greater than mine, the answer is the same as it would be if my number were two more than it is."

What is Sir Bale's number?

Answer on page 77.

INSPECTING THE TROOPS

An officer on horseback rides slowly down a line of sixty mounted troops placed 10 feet apart. Beginning with the first man, the officer takes 29 seconds to reach the thirtieth man.

At that rate, how long will it take him to reach the sixtieth (last) man?

Answer on page 76.

HOW MUCH FOR A KNIFE, SWORD, OR ARROW?

In Pymm, thirty-six coins will buy one knife, one sword, and nine arrows. Two swords can be traded for one knife and four arrows.

What is the price for each item purchased separately?

Answer on page 75.

STRANGE RABBITS

Rabbits in Pymm are not like rabbits elsewhere. They can only jump two distances: 5 feet or 7 feet, either forward or backward. To reach an object 12 feet in front of it, a Pymmian rabbit would take one 5-foot jump and one 7-foot jump. This being the case, a Pymmian rabbit has to do some mathematical thinking in order to reach a carrot growing 13 feet ahead of it.

Assuming that the rabbit moves in straight-line jumps toward the carrot, what is the fewest number of jumps that will take the rabbit to a carrot 13 feet ahead of it?

Answer on page 82.

WHAT IS THE HUMAN POPULATION OF SOUTH PYMM?

One-third, one-fourth, one-fifth, and one-seventh of the human population of North Pymm, which has fewer than five hundred human inhabitants, are all whole numbers, and their sum is exactly the human population of South Pymm. What is the human population of South Pymm?

Answer on page 91.

TRICKY SQUARES

Dob asked Gregory, "Is it possible to color these nine dots either white or black so that in every possible square the diagonally opposite dots are different in color?" (See illustration.)

Gregory proceeded to show Dob how he thought it might be done. Dob then argued that Gregory's solution was not a solution. Who was right? Explain by proving how the requirements either can or cannot be met.

Answer on page 89.

MEASURING TWO GALLONS OF CIDER

"I want 2 gallons of cider for me and my pals," said Mongo to the pub owner.

The pub owner replied, "I have a 3-gallon container and a 4-gallon container. Will you use one of those and guess at the amount?"

"I don't need to guess," said Mongo. "I can measure exactly 2 gallons using the containers you have."

How can Mongo do this?

Answer on page 77.

HOW MANY DWARVES?

An outdoor amphitheater holds 120 humans or 144 dwarves. If 90 humans are in the amphitheater, how many dwarves can also be admitted?

Answer on page 72.

CHAPTER II

WIZARDS, DRAGONS, AND OTHER MONSTERS OF PYMM

CHASED BY A GLUB*

Phipos was 5 feet away from a glub when he saw the fat, swollen monster advancing toward him. Phipos knew that he and the glub ran at the same rate and walked at the same rate (which was, of course, slower than their running rate), but Phipos also knew that the glub could spray him with an irritating fluid from almost 5 feet away. So he immediately began to run toward the safety of a fort several miles in the distance. At that same instant, the glub started chasing after Phipos.

Curiously, Phipos reached safety by first running half the *time* it took him to cover the distance to the fort, then walking half the time. The glub, which ran the first half of the *distance* to the fort and walked the other half, was never closer to Phipos than the original 5 feet. Can you explain this?

Answer on page 67.

A DRAGON STORY*

One day at a pub in Athey two Knights of the Golden Sword, the bravest knights in all of Pymm, swapped dragon stories.

One said, "I saw two dragons together just outside their cave. Apparently one was the mother of the other, which wasn't yet fully grown. The body of the mother dragon was around three times as long as her head, while her head was approximately twice as long as the head of her offspring and about as long as her own body minus her own tail. The body of the younger dragon was about 2 feet longer than its head, while its tail was about one-third the length of the tail of its mother. The length of both dragons together was about 48 feet."

Ignoring the words "around," "approximately," and "about," what were the lengths of the head, body, and tail of the mother dragon and her offspring?

Hint on page 56.
Answer on page 63.

HOW FAR APART WERE THE DRAGONS?

The dragons Argothel and Bargothel like to get together for fiery conversations. They live some distance apart, each in his own cave. One day Argothel left home to visit Bargothel at exactly the same time that Bargothel left home to visit Argothel. The day being most agreeable, both dragons decided to proceed at a rather leisurely rate—for dragons. So, rather than fly, they walked. Argothel walked at a constant rate of 24 miles per hour and Bargothel at a constant rate of 36 miles per hour.

How far apart were they 5 minutes before they met?

Answer on page 70.

THE FARMER AND THE HOBGOBLIN*

A farmer was at work clearing a rocky field for planting when an ugly hobgoblin approached him. The hobgoblin promised the farmer great wealth if he followed these instructions:

"Pick up a stone and carry it to the other side of the field. Leave the stone there and pick up another stone and drop it on this side of the field. Continue doing this, and each time you cross the field and return, I will double the number of coins you have. However, because I don't want you to be overly rich, you must pay me sixteen coins each time after I have doubled your money."

The farmer, thinking only of the doubling of his money, readily accepted. Even after giving sixteen coins to the hobgoblin following his first crossing, he did not take the time to figure out what would happen if he continued with the hobgoblin's bidding.

After four crossings, the farmer was not only too tired to move, but he also had to give the hobgoblin his last sixteen coins. The farmer's pockets were empty and the hobgoblin went away laughing.

How much money did the farmer start with?

Hint on page 56.
Answer on pages 83–84.

THE WIZARDS' DRAGON SPELLS

Each of three towns in Pymm is menaced by a dragon living in a cave above the town. The wizards Malefano, Sagareth, and Thaumater created these three dragons. Sagareth's dragon will menace one of the towns for the same number of years as the square root of the number of years of Thaumater's dragon curse on a second town. Sagareth's dragon curse will also last the number of years equal to half the square root of the number of years of Malefano's dragon curse on the third town. Thaumater's dragon curse will last the number of years equal to twice the square root of the number of years of Malefano's curse. How long will each curse last?

Hint on page 56.
Answer on page 88.

WIZARD RANKINGS

Every Saturday at Pymm's End, a peninsula at the tip of southwest Pymm, there is a ranking of Pymm's End's six resident wizards. The rankings are based on the wizards' feats of magic for the previous six days. The highest ranking is 1 and the lowest is 6.

The rankings published on the Saturday before last listed Alchemerion as the highest because he had cast a spell that imprisoned an evil dragon inside an iceberg. The others were: 2. Bogara, 3. Chameleoner, 4. Deviner, 5. Elvira, 6. Fortuna.

Last Saturday's rankings had each wizard ranked in a different position from that of the previous Saturday. The following facts are known:

1. Bogara's change in ranking was the greatest of the six.
2. The product of Deviner's rankings for the two weeks was the same as the product of Fortuna's ranking for the two weeks.

What were the new rankings?

Hint on page 56.
Answer on pages 93–94.

HOW LONG ARE THE GLUBS?

Five glubs that tried to enter the town of Galvinchy were slain by ten Knights of the Golden Sword. The knights laid the five glubs' bodies tail to head with 5 feet between each tail and the next head. The five glubs covered 200 feet. The first, third, and fifth glubs were all the same length, as were the second and fourth glubs. Each glub was either 10 feet longer or 10 feet shorter than its neighboring glub. Furthermore, each glub's length was a multiple of 10.

What was the length of each glub?

Answer on page 72.

DID THE DRAGON CATCH PRYOR?

The dragon Wivere smelled the half-elf Pryor at the same time that Pryor, who had been hunting in the forest, became aware of smoke and fire rising from the direction of the dragon's mountain cave. Pryor realized the dragon was active and might try to catch him. Knowing that Wivere was afraid of water, he began to run toward the seashore.

Wivere smelled Pryor but hesitated for 6 seconds before he began to run after the half-elf. The dragon ran because his wings were undeveloped compared to the larger ones of his ancestors.

Wivere was 5 miles directly north of Pryor when Pryor began to run toward the sea 2 miles directly to his south. Pryor, who could run much faster than a human, ran at a rate of 20 miles per hour over the 2-mile distance.

Wivere's speed, however, was not constant. He ran the first mile at a rate of 20 mph, the second mile at a rate of 40 mph, the third at a rate of 80 mph, the fourth at a rate of

160 mph, and so forth—doubling his speed after running each mile. Did Pryor make it to the safety of the sea?

Answer on page 93.

HOW OLD IS THE WIZARD ALCHEMERION?

"How old are you, Alchemerion?" asked one of the wizard's apprentices.

The wizard answered with a riddle, "I am still very young as wizards go. I am only three times my son's age. My father is 40 years more than twice my age. Together the three of us are a mere 1240 years old."

How old is Alchemerion?

Answer on page 75.

MINOTAUR FIGHTERS

Three Minotaur leaders lent each other fighters from their armies. First, Logi lent Magnus and Nepo as many fighters as each already had. Later, Magnus lent Logi and Nepo as many fighters as each already had. Still later, Nepo lent Logi and Magnus as many fighters as each already had. Each leader then had forty-eight fighters. How many fighters did each have originally?

Hint on page 56.
Answer on page 78.

CONTESTS OF SKILL

A tournament in Pymm pitted each elf contestant against every knight and Minotaur contestant, each knight contestant against every elf and Minotaur contestant, and each Minotaur contestant against every elf and knight contestant in three contests of skill: wrestling, archery, and knife throwing. In other words, each type of being—elf, knight, Minotaur—competed against everyone in the other groups of beings in three different contests.

These facts are known:

1. The number of Minotaurs in the tournament was one fewer than the number of elves.
2. Nine of the contestants were either knights or Minotaurs.
3. The number of elf vs. knight competitions in the three skills was seventy-two; there were more knights than elves.
4. Twenty-seven contests were held each day.

How many days did the tournament last?

Hint on page 56.
Answer on page 67.

HUMAN VS. MINOTAUR

A human warrior must fight seven competitors with his lance. Six of the seven are humans, and one is a huge Minotaur with a fierce reputation. The warrior is certain he will lose the battle with the Minotaur. He is about to end his life quickly by submitting to the Minotaur's lance, when the Minotaur issues this challenge:

"Arrange us seven competitors in a circle," he says. "Choose a competitor from this circle and, moving clockwise and counting that competitor as the first position, count to the seventh warrior and fight him. From there, assuming you win the contest, count seven more places clockwise. Continue in this way until you reach my place on the circle. If you manage to defeat all the others before your turn comes to fight me, I will spare your life."

How can the human warrior make sure the Minotaur is his seventh competitor?

Answer on page 76.

A WIZARD COMPETITION*

Although very fond of one another, the wizards Alchemerion and Elvira often engage in fierce competitions of magic. One day, the two strolled arm in arm through their garden where they saw butterflies in colorful abundance among the flowers. Suddenly, the butterflies began to change into handsome men.

"At it again, are you, Elvira!" Alchemerion shouted, as he began to transform butterflies into beautiful princesses just as fast as he could.

When all the butterflies had been transformed, there were more handsome men than beautiful princesses.

"Not fair! You had a head start, Elvira," Alchemerion shouted, and, without warning, he began to change the handsome men into beautiful princesses.

Elvira's answer to her rival was to begin transforming the beautiful princesses into handsome men. When all the transforming was done in the second stage of their competition, the number of handsome men was the same as the number of beautiful princesses. One-third of the butterflies originally transformed into beautiful princesses had been changed to handsome men by Elvira. Two-fifths of the butterflies originally transformed into handsome men had been changed to beautiful princesses by Alchemerion.

"I won," Alchemerion declared, but the words had hardly passed his lips when Elvira answered, "Watch this, you old mage," and once again she began to change the beautiful princesses into handsome men.

"You witch, you!" yelled Alchemerion while he worked his magic as fast as he could. But finally he surrendered. Elvira had won again.

During the last stage of the contest, three-fourths of those who were originally changed from handsome young men to beautiful princesses were changed back to handsome young men, while half of those who were originally changed from beautiful princesses to handsome young men were changed back to beautiful princesses. As a result there were twelve more handsome young men than beautiful princesses.

There were no double transformations during any stage of this competition. In each of the second and third stages, Alchemerion transformed only those who had been changed to handsome men by Elvira during the previous stage and Elvira transformed only those who had been changed to beautiful princesses by Alchemerion during the previous stage. How many butterflies were there originally?

Hint on page 56.

Answer on pages 64–65.

CHAPTER III

EVERYDAY LIFE IN PYMM

GOING TO THE FAIR

One bright summer morning the brothers Bopp and Clack left their hut on horseback to go to a fair some distance away. The two men left at the same time and arrived at the fair at the same time, 12 hours after they began the trip. Clack had a fast mount, while Bopp had a slow, old nag. Both brothers stopped to rest at various times as they rode toward their destination. However, Clack, who rode twice as fast as Bopp during the trip, rested often so Bopp could catch up. In all, Clack rested for as long as Bopp rode, and Bopp rested for as long as Clack rode.

What were the riding and resting times for each man?

Hint on page 56.
Answer on pages 68–69.

HOW MANY CAKES?

A banquet was given to celebrate a truce in East Pymm that put an end to clashes between humans and elves. Citizens from all races of East Pymm—humans, elves, dwarves, and half-elves—came together at the banquet.

After a huge dinner, sweet cakes were served for dessert. The first four beings at the banquet table emptied the first plate of cakes. (Luckily the castle cook had made enough sweet cakes for everyone to eat.) Heartnik, a human, took one-fourth of the cakes on the plate. Scowler, a dwarf, took one-third of the remaining cakes. Then Goodin, an elf, helped himself to half of the cakes left. Finally, Loglob, a half-elf, ate six cakes—all that were left on the plate.

How many cakes were on the plate in the beginning, and how many did each being take?

Answer on page 72.

WINNING AT MARBLES

Dravid and Gopar each had a certain number of playing marbles. Dravid lost a game to Gopar and had to give Gopar half of his marbles. In the second game, Gopar lost and had to give three-fourths of his marbles to Dravid, who now had thirty marbles. Finally, the two played a third game. Gopar won and acquired some of Dravid's marbles.

At this point, each boy had exactly the number of marbles he had started with, and Dravid had twice as many marbles as Gopar. How many marbles did each have in the beginning?

Answer on page 93.

THE FOOLS' SHOW

The promise of buffoonery and clever wit brought thirty residents of Athey to a fools' show that came to their village once a year. The ticket buyers included children, knights, and other adults. Most of the children sneaked in without paying, making the number of "other" adults who paid more than the number of children who paid but not as large as twice the number of children who paid. Fewer knights paid than children.

Children paid 5 stickels each for admission; knights paid 25 stickels each, and all other adults paid 50 stickels each. Total admission receipts were 1000 stickels. How many children, how many knights, and how many other adults paid for admission to the show?

Hint on pages 56–57.
Answer on pages 84–85.

TO THE KING'S CASTLE*

All the villagers in the realm of King Althur were invited to a Christmas gala held at the king's castle. Alf and his wife, Beryl, left their cottage at 11 A.M. that morning and traveled by ox cart to the castle. At 11:30, an eager and impatient Beryl asked Alf, who was driving at a fast pace, "How far have we gone, dear?"

Alf replied, "Three times as far as the distance from here to the inn, where I will stop and have some nourishment."

The couple stopped at the inn for food and departed at 1 P.M. to continue their journey. Having overeaten, Alf became sleepy and drove much slower than before their stop at the inn. Afraid of being late, Beryl asked her husband, at 2:00 P.M., exactly 2 miles from the point where she had asked her first question, "Do we have much farther to go, Alf?"

Alf mumbled, "Three times as far as we have come since leaving the inn."

Beryl then demanded that she take over the driving, to which a sleepy Alf readily agreed. The couple finally arrived at the castle at 3:30 in the afternoon. Despite the long journey, both thoroughly enjoyed all the festivities. What was the distance from the couple's cottage to the castle?

Hint on page 57.
Answer on pages 88–89.

BUILDING A BRIDGE

The dwarves Dobbit and Mobbit are building a bridge over a narrow stream. Dobbit can do the job alone in 30 hours; Mobbit can do the job alone in 45 hours. Dobbit worked on the project alone for 5 hours before Mobbit joined him. The two then finished the job together.

How long did it take the two to finish the job that Dobbit had started alone?

Hint on page 57.
Answer on page 66.

SHARING A JOB

Abelard and Brendan can do a job together in 10 days. Abelard and Cullen can do it together in 15 days. Brendan and Cullen require 30 days to do the job together. How many days would it take each to do the job if he worked alone?

Answer on page 81.

A JOB NOT DONE ON TIME*

Four cave dwarves, each of whom works at the same rate, were to complete a mining job according to a schedule. However, because of an argument, two of the four quit after working one day. The remaining two dwarves finished the job, but it required two more days than had originally been scheduled. How many days were originally scheduled for completion of the job?

Hint on page 57.
Answer on pages 63–64.

AT WHAT TIME DOES THE WAGON DRIVER LEAVE HIS HUT?

Once a week a wagon driver leaves his hut and drives his oxen to the river dock to pick up supplies for his town. At 4:05 P.M., one-fifth of the way to the dock, he passes the smithy. At 4:15 P.M., one-third of the way, he passes the miller's hut. At what time does he leave his home? At what time does he reach the dock?

Hint on page 57.
Answer on page 66.

MEETING THE STONE CUTTER

Every morning, a cart driver leaves the stone quarry to drive to the ferry landing, where he picks up an arriving stone cutter. The driver arrives at the landing at 6:00 A.M. and takes the stone cutter back to the quarry.

One morning, the stone cutter woke up before his usual time, took an early ferry, and, once across the river, began walking to the quarry. The cart driver left the quarry at his usual time and met the stone cutter along the road to the ferry landing. He picked up the stone cutter and took him the rest of the way to the quarry. The stone cutter arrived 20 minutes earlier than usual. At what time did the cart driver meet the stone cutter?

Hint on page 58.
Answer on page 77.

HOW EARLY WAS THE BARGE?

Every day a cart is sent from a village to meet a barge at the river dock. One day the barge arrived early, and the cargo normally picked up by the cart was immediately sent toward the village by horse. The cart driver left the village at the usual time and met the rider along the way, after the rider had traveled for 8 minutes. The rider handed the load to the cart driver, who went back to his village, arriving home 24 minutes earlier than usual.

How many minutes early was the barge?

Hint on page 58.
Answer on page 69.

HOW MANY HANDSHAKES?

Fifteen knights were invited to a sumptuous meal at the castle in Belmar. Before sitting down, each of the fifteen knights shook hands with each of the other knights.

How many handshakes occurred?

Answer on page 73.

MUGS CLINKING

At the same banquet, after shaking hands, each of the fifteen knights sat at the round table and clinked his mug with the Knights to his immediate left and immediate right.

How many times did mugs clink?

Answer on page 80.

MORE MUG CLINKING*

King Wincher's knight Algar told this story to a friend.

"I recently attended a gathering with three others of King Wincher's knights: Cautious, Elveron, and Grace, and four of King Dracon's knights: Bolder, Daring, Fanciful, and Haldar. Each of the eight knights found that a mutual archenemy from the other kingdom was at this gathering, and altogether there were four such duos, with no knight being in more than one duo.

"Many drinking mugs were clinked that night, but no knight clinked mugs with his archenemy. Afterwards, I asked each of the seven other knights how many mugs he had clinked with his, and to my surprise each had clinked a different number of mugs.

"The only knight who serves King Dracon with whom I clinked mugs was Daring, whose archenemy is Cautious. The total of the number of mugs clinked by the other three

Wincher knights was five more than the total of the number of mugs clinked by Dracon's four knights. Note that a mug clinking between two knights counts as a mug clinking for each. For example, since I clinked mugs with 'X,' that counts as a mug clinking for both me and 'X.'

"How many and whose mugs did my archenemy clink? Also, how many mugs did Sir Daring clink?"

Hint on page 58.
Answer on pages 79–80.

THE SHOPPING TRIP

Marcella and her husband, Justice, went into town to shop. Marcella bought a robe for herself, while Justice bought a chest and a mantle for himself. All together, the three items cost 80 stickels.

On their way home, Marcella teased Justice, "If you hadn't gotten that mantle, there would have been exactly enough money to buy me that bracelet I liked. Of course, that wouldn't have been fair as it would have meant that the amount spent on me was just 10 stickels short of being five times the amount spent on you. As it was, exactly the same amount was spent on each of us." What was the price of each of the purchased items?

Answer on pages 87–88.

THE JEWEL CHEST

Amarina has a jewel chest containing rings, earrings, and pins. The chest contains twenty-six pieces of jewelry, with each pair of earrings counting as two pieces. Amarina has 2½ times as many rings as pins, and the number of pairs of earrings she has is 4 less than the number of rings she has. Half of her earring collection was made by a certain artisan while the other half was made by another.

A huge breeze blew out the candle in her room one night, and Amarina was forced for reach for her jewelry in the dark. How many pieces of jewelry did she have to remove from the chest to be certain she had a matching pair of earrings? Assume Amarina can't tell a ring from an earring from a pin in the dark, and that any two earrings from one artisan match.

Answer on pages 85–86.

HOW MANY SCHLOCKELS?

Altus says to Bott, "Can you figure out how many schlockels I have in my pockets?" He then gives Bott three clues:

1. If the number of schlockels I have is a multiple of 5, it is a number between 1 and 19.
2. If the number of schlockels I have is not a multiple of 8, it is a number between 20 and 29.
3. If the number of schlockels I have is not a multiple of 10, it is a number between 30 and 39.

How many schlockels does Altus have in his pockets?

Answer on page 74.

A ROUND TABLE ARRANGEMENT

The brothers Bob, Cob, Hob, Lob, Rob, and Tob always take the same seat when they have their meals at their round dinner table. The following diagram indicates their seating. The following is known:

1. Lob's seat is separated from Bob's seat by exactly one of the other brothers.
2. Hob's seat number differs from Cob's and Lob's, positively or negatively, by 2 and 5 in one order or the other.
3. Cob's number is 1 larger than Rob's number.
4. Bob's number is either 1 larger or 1 smaller than Tob's number.

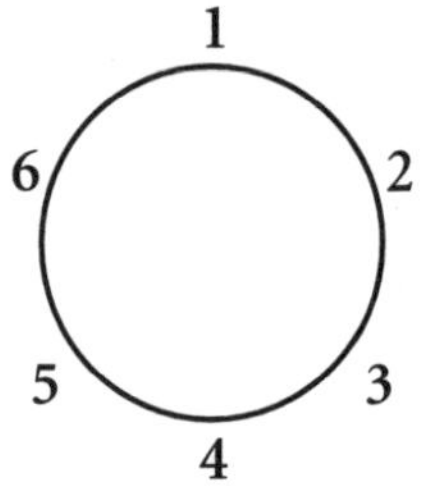

Which brother sits in which seat?

Answer on page 64.

HOW MANY EGGS?

Each morning, a farm woman collects the eggs her hens have laid. One day, she stumbled as she left the coop and all the eggs were broken.

"How many eggs did you collect?" asked her daughter.

"I don't know," said the woman, "but I do remember that when I divided the number of eggs by 2, there was one egg left, when I divided the number by 3 there were no eggs left, and when I divided by 5, there were three eggs left."

The woman has more than four eggs and fewer than forty. How many eggs were broken?

Hint on page 58.
Answer on page 73.

CHAPTER IV

MORE PUZZLES

PERFORCE ARRIVES TOO LATE

King Perforce led his army from his castle toward Halsen to free that town from its captor, the evil Malfor. He had planned the march to Halsen with great care but his army was delayed three times in its march. First, torrential rains slowed down the infantry for half a day. Next, when they arrived at the town of Aker, Perforce found that the bridge connecting Aker to the next town had been washed out. The extra time required to find a place to safely ford the river and then cross it was one-tenth the total number of days originally planned for the entire march.

After crossing the river, the troops were so fatigued that Perforce was obliged to halt for a three-day rest. When the army finally arrived at Halsen they found the town abandoned and Malfor nowhere to be found. The march to Halsen had, altogether, taken one-third more days than originally planned. How many days had been planned for the march to Halsen and how many days did it actually take?

Answer on page 81.

SWITCHING ALLEGIANCE

Both Lavar and Malcavar have small troops of volunteer soldiers at their command. Together the two have more than ten but fewer than thirty soldiers. One day, one of the soldiers, Colin, decided to leave Malcavar's troop and join Lavar's. This had the effect of making the number of soldiers in each troop the same. Eventually, Colin rejoined Malcavar's troop, but then Draal decided to give his allegiance to Malcavar rather than Laval. Once Draal switched allegiance, both Laval and Malcavar had a prime number of soldiers. How many soldiers did each then have?

Answer on pages 82–83.

ABLE AND GALLANT SOLDIERS

Both Sir Able and Sir Gallant, Knights of the Golden Sword, have small volunteer armies at their disposal. Some of the soldiers are foot soldiers, some are mounted troops. Together, the two armies total fifty men. The number of mounted troops in Gallant's force equals the number of foot soldiers in Able's force. Sir Able has two fewer troops in total than Sir Gallant, and the number of mounted troops in Able's army is four fewer than the number of mounted troops in Gallant's army.

How many mounted troops and foot soldiers does each knight have?

Answer on pages 65–66.

MORE JOUSTERS

The pairs-jousting tournament was about to begin at King Althur's castle when some unknown knights appeared and asked that they be allowed to fight in the competition. In the tournament, each knight would compete against every other knight unless a serious injury should require a withdrawal. The king decided that the unknown knights would be allowed to compete, which meant that 26 more pairs-competitions had to be scheduled.

How many knights were to compete in the joust before the unknown knights asked to be added to the list of competitors? How many unknown knights were there?

Hint on page 58.
Answer on pages 78–79.

THE SONS OF BLYTHE

The noblewoman Blythe said to the noblewoman Alexis, "I have three sons. They are all less than 10 years in age, and the product of the ages of the two youngest equals the age of the one who, in years, has the greatest age. How old are they?"

"I can't determine their ages from that information," said Alexis.

Blythe added, "My final clue is that the sum of their ages is a prime number." How old are Blythe's sons?

Answer on page 88.

THE DAUGHTERS OF ALEXIS

After Alexis had figured out the ages of Blythe's sons, Alexis herself, who has three daughters, asked Blythe, to attempt to figure out their ages.

"Your first clue is that the sum of their ages is 11," said Alexis.

"That is not enough information," replied Blythe.

"The product of their ages is either 16 years less or 16 years more than your age," added Alexis.

"Still not enough information," said Blythe after careful thought.

"The daughter whose age, in years, is the greatest is learning to play chess," said Alexis.

Blythe was then immediately able to determine the ages of Alexis's three daughters. What are their ages?

Hint on page 58.
Answer on page 83.

WHO RODE FASTER?

Two Knights of the Golden Sword, Sir Pure and Sir Chaste, started on horseback at the same time from the same place to take the same route to a town that was threatened by a dragon.

Pure rode three times as long as Chaste rested on the trip, and Chaste rode four times as long as Pure rested. The two arrived at their destination at the same time. Who rode faster?

Answer on page 92.

HOW FAR FROM CASTLETON TO DEVIL'S PEAK?

A horseman left Castleton and rode toward Devil's Peak at the same time that a second horseman left Devil's Peak and rode toward Castleton using the same route. The two met for the first time at the point on the route that is four miles from Castleton.

The horsemen continued riding to their destinations, then turned and rode back toward home at the same rate. Again they met, this time two miles from Devil's Peak.

Assuming both riders traveled at a constant speed, what is the distance between Castleton and Devil's Peak and what was the relative speed of the two riders?

Hint on page 58.
Answer on pages 70–71.

HOW DID THE ARCHERS CROSS THE RIVER?

A company of archers had to cross a river that had no bridge across it and was too deep to ford. They had just about given up on finding a way to do this when one of them saw two children on the river in a small rowboat. The boat was large enough to hold the two children or one archer, but was too small to hold two archers, or even one child and one archer. How did the archers get across the river?

Answer on page 69.

HOW CAN EVERYONE CROSS THE RIVER?

A half-elf, an elf, a human, and a dwarf have to cross a river from the town of Ak to the town of Tok. They have a boat that will hold any two of the four, but no more.

Long-standing animosities exist between certain of these four beings, which make it undesirable for certain ones of the four to be together as pairs. While the half-elf gets along with the other three beings, the elf dislikes the human. The human dislikes the dwarf. The elf and the dwarf like and respect each other. How can the four beings cross the river if only the half-elf does the rowing, and no uncongenial pairs are together alone?

Answer on page 69.

HOW FAST DOES LUCIEN'S HORSE NEED TO RUN?

Mnemo's horse is so fast that it runs four times as fast as Wister's old nag. Lucien also has a horse, but no one knows how fast it can run. Even so, Mnemo believes that his own horse is faster, and Wister's is slower, than Lucien's horse. So, when the three decide to run a 2-mile race, they agree that in an attempt to have a fair race, Lucien's horse will run the entire 2-mile course, while Mnemo's and Wister's horses will each race for only a mile. Mnemo's horse will run the first mile and Wister's horse will run the second.

Compared to Mnemo's horse, how fast does Lucien's horse need to run in order to finish the race at the same time as Wister's?

Hint on page 58.
Answer on page 71.

THE PRICE OF CANDY*

Gwendar, a cave dwarf, loves sweets. When a candy maker came into town, she was one of his first customers. She purchased three different kinds of candy: bonbons, sweetles, and chocos, buying as many pieces of each kind of candy as its price per piece in shukels.

Gwendar paid an average of 7 shukels for each piece of candy, but each of the three kinds cost a different amount of shukels per piece. Bonbons cost the most, sweetles cost less, and chocos cost the least per piece. All of the candies cost at least 4 shukels per piece. Compared to bonbons, one of the other two kinds of candy costs 3 shukels less per piece. Gwendar spent less than 150 shukels. How many of each kind did she buy?

Hint on page 59.
Answer on pages 86–87.

FEEDING THE HORSES

Distal had sufficient hay and corn to feed his six horses for only 30 more days of the harsh winter, not enough for the remaining 75 days before spring arrived. On the seventh day, before feeding time, Distal sold four of his horses. Will he be able to feed his remaining two horses for the rest of the winter?

Answer on page 68.

THE ISLAND OF ODDS

On the island of Odds, off the coast of Pymm, one-third of the native people always lie, one-third always tell the truth, and one-third are "normals" (people who sometimes tell the truth and sometimes lie). The chances of encountering any one of the three natives on a road on the island are the same. If a traveler meets a native of Odds on the road on each of two successive days, what is the probability that at least one of the two natives is a "normal"?

Answer on page 85.

WHICH COIN IS LIGHTER?

Peppi, Bogara the Wizard's newest apprentice, groaned as he told Welch, an elder apprentice, of the puzzle Bogara gave him.

"I've got fifty coins here. Fifty! Can you believe that they all weigh the same, except for one that's a tiny bit lighter than the other forty-nine. And I've got to find the lighter coin in four weighings on our balance scale. How in the world can I do that?"

Patting Peppi on the back, Welch told him that, indeed, the problem could be solved with a maximum of four weighings. Explain how this can be done.

Hint on page 59.
Answer on pages 91–92.

TRINKETS TO SHARE

After Sir Elveron and Sir Grace, Knights of the Golden Sword, had slain the dragon Wyngard, they found a great treasure in Wyngard's cave. The treasure included a number of small trinkets that the knights gave to a group of fewer than ten girls. The girls found that they were able to divide the trinkets equally among themselves.

After this division had been done, Moira, one of the girls, suggested that it would be more equitable to divide the trinkets by families rather than by individuals. Among the girls, there were two groups with two sisters (Moira was not in either group). The rest of the girls were unrelated to each other. A redivision by families would have meant that the trinkets per family were five more than the trinkets per girl.

The girls argued among themselves over this way of dividing the trinkets. The two sets of sisters were especially insistent on retaining the division-by-individuals plan, because each sister would get more trinkets using this plan. Before a final decision was made, Felice, one of the girls, decided that she did not want any trinkets. Her share was then divided among the other girls and the result was that each now had more trinkets, but still all had an equal share. Moira then decided to withdraw her suggestion of dividing the trinkets by families. She was satisfied with this new state of things. How many trinkets did each girl end up with and how many girls shared in the division?

Hint on page 59.
Answer on page 90.

A TIGER OR A TREASURE*

A young man named Ivan was wandering in a land East of Pymm in search of adventure when he was captured by the much-feared warrior leader Kusan. Kusan had a reputation for showing no mercy toward his victims, but Ivan was somewhat relieved when Kusan informed him that he would be given a chance to save his life and even acquire a treasure. It seems that Kusan was fascinated by math and probability games. To amuse himself, he posed this problem for the captured Ivan. Kusan lead the young man to three caves, each secured by a barricade.

He said, "Choose a cave, Ivan. In two of them are tigers that would like nothing better than to have you for dinner. But in one cave lies a treasure beyond your sweetest dreams. If you choose this cave, not only will I spare your life, but the treasure is yours. After you choose a door, I will open one of the others to show you a tiger, and give you a chance to change your mind."

Shivering in his boots, Ivan closed his eyes and pointed to a cave that we will call Cave A. When Ivan had made this choice, Kusan, who knew the contents of all three caves, had his soldiers remove the barricade from one of the caves not chosen by Ivan to show him that it held a tiger.

Kusan then said to Ivan, "If you wish to change your selection, you may do so now."

Ivan thought for several minutes before he decided to change his mind and choose the other cave. Should he have done so? If yes, why? If no, why not?

Hint on page 59.
Answer on pages 62–63.

DRAWING STONES

Gopar put some black stones and some white stones into a jar. He then asked his friend Jason to reach into the jar and take out a stone. Jason drew out a black stone. Gopar asked Jason to draw out another stone, and, once again, Jason drew out a black stone.

"There must have been more black than white stones in the jar," said Jason. "I wonder what the probability is of me drawing a black stone on a third try?"

Gopar replied, "Exactly nine-tenths of what it was of drawing a black stone on your first draw."

Gopar told Jason that he had put "ten, give or take two or three" stones into the jar. Jason was then able to determine how many stones of each color had been in the container in the beginning. Can you do the same?

Hint on page 59.
Answer on pages 67–68.

HINTS

A Dragon Story: Set up six equations with six unknowns: mother dragon's head, body, tail; younger dragon's head, body, tail. Five of the unknowns can be expressed in terms of the sixth. So although the puzzle sounds difficult, it really isn't.

The Farmer and the Hobgoblin: Figure out how many coins the farmer had just before the *final* doubling.

The Wizards' Dragon Spells: Say M, S, and T are the number of years the curses will last as cast by Malefano, Sagareth, and Thaumater, respectively. Write three equations and solve them: one in S and T, a second in S and M, a third in T and M.

Wizard Rankings: From statement (1), there are two possibilities regarding the ranking changes of Fortuna and Deviner. One of these leads to a contradiction of another clue.

Minotaur Fighters: Start at the end and work toward the first lending.

Contests of Skill: Say E, M, and K = the number of elves, Minotaurs, and knights respectively. Write three equations: one in M and E, a second in K and M, a third in E and K.

A Wizard Competition: Solving this puzzle involves setting up and solving a system of equations.

Going to the Fair: If b = the length of time Bopp rode (and Clack rested) and c the length of time Bopp rested (and Clack rode), then $b + c = 12$.

The Fools' Show: Figure out what integer can be divided into the number of children's tickets sold. This limits the number of possible answers to the question of how many children bought tickets to only a few. Consider each possible answer until you find the one that meets all the givens in the puzzle. Once you find the number of children, write and solve a system of equations to find the number of knights and other adults who bought tickets.

To the King's Castle: Draw a diagram.

Building a Bridge: First figure out what part of the job Dobbit and Mobbit do in *one* hour when they work together.

A Job Not Done on Time: If x is the part of the total job done by one dwarf in one day, then 4x is the part of the job done by four dwarves in one day, and 2x is the part of the job done by two dwarves in one day.

At What Time Does the Wagon Driver Leave His Hut?: First figure out what part of the total trip the driver makes in the 10 minutes from 4:05 to 4:15.

Meeting the Stone Cutter: The cart driver spent 20 minutes less time traveling than he usually did. Of this 20 minutes, half would have been spent going toward the ferry and half coming from the ferry.

How Early Was the Barge?: If the cart driver had not met the rider, he would have needed 12 more minutes of time to get to the dock at the usual time.

More Mug Clinking: One knight clinked six mugs, another clinked five, a third clinked four, and so on. Try making a diagram.

How Many Eggs?: The number of eggs broken is an odd number. Make use of the facts that the number of eggs is less than 40 and equal to $5k + 3$, for some positive integer, k. Then make use of the fact that the number is evenly divisible by 3.

More Jousters: If x = the number of original competitors, then $[x(x - 1)] \div 2$ = the number of jousts. Figure out how many jousts there would be if there were y more competitors.

The Daughters of Alexis: Ten different combinations of three numbers add to 11, so the first clue is insufficient. You need to figure out why the next clue still fails to give sufficient information to Blythe.

How Far from Castleton to Devil's Peak?: Draw a diagram showing each rider's route.

How Fast Does Lucien's Horse Need to Run?: If r = the rate of Mnemo's horse in miles per hour, then the rate of Wister's horse is $(1/4)r$.

The Price of Candy: Say x = the number of bonbons purchased. If y and z = the number of sweetles and chocos purchased, in one or the other order, then the information about the total cost leads to the equation $7(x + y + z) = x^2 + y^2 + z^2$. Solving this puzzle requires a bit of trial and error.

Which Coin Is Lighter?: Divide the fifty coins into three groups: two with seventeen coins each and one with sixteen coins. The first weighing should compare the two groups of seventeen coins each.

Trinkets to Share: If t = the number of trinkets and g = the number of girls, then t is evenly divisible by g as well as by the number of families and the number of sharers after Felice's share was eliminated.

A Tiger or a Treasure: Make use of the fact that Kusan knew the contents of all three caves and showed Ivan a cave containing a tiger.

Drawing Stones: If the number of black stones in a container is b and the number of white stones in the same container is w, then the probability of drawing a black stone is $b/(b + w)$.

ANSWERS

A Tiger or a Treasure: Ivan made the more informed decision, although, of course, we do not know the result. Switching his choice of caves made the probability of Ivan's winning the treasure equal two-thirds. Remember that Kusan knew the contents of all three caves. Regardless of Ivan's first choice, Kusan said he would open up a cave that contained a tiger to show Ivan.

The puzzle refers to Ivan's first choice as Cave A. The other two caves will be known as Cave B and Cave C. If Cave A was the one with the treasure, then Kusan would have opened either Cave B or Cave C, and Ivan would have switched his choice to the one of those two that Kusan didn't open; in this case Ivan would have lost. If Cave B held the treasure, then Kusan would have opened Cave C; if Ivan switched to B, he would have won the treasure. If Cave C held the treasure, then Kusan would have opened Cave B; if Ivan switched to C, he would have won the treasure. Assuming Ivan initially chose Cave A, here are the results in a table.

CAVE CONTENTS

Cave A	Cave B	Cave C
Treasure	Tiger	Tiger
(Kusan opens either Cave B or C. If Ivan switches, he loses. If he doesn't switch, he wins.)		
Tiger	Treasure	Tiger
(Kusan opens Cave C. If Ivan switches, he wins. If he doesn't switch, he loses.)		
Tiger	Tiger	Treasure
(Kusan opens Cave B. If Ivan switches, he wins. If he doesn't switch, he loses.)		

By changing his selection, Ivan put himself into the position of having two out of three chances of winning the treasure and one out of three of finding a tiger that would eat him. The reverse would have been true if he had not changed his selection.

A Dragon Story: Mother: head, 6 feet; body, 18 feet; tail, 12 feet. Young dragon: head, 3 feet; body, 5 feet; tail 4 feet.

Say r, s, and t are the lengths of the mother dragon's head, body, and tail, respectively. Say v, w, and x are the lengths of the younger dragon's head, body, and tail, respectively. Six equations in the six unknowns can be derived from the given information: (1) $s = 3r$; (2) $r = 2v$; (3) $r = s - t$; (4) $w = v + 2$; (5) $x = (1/3)t$, or $t = 3x$; (6) $r + s + t + v + w + x = 48$.

Expressing each of s, t, v, w, and x in terms of r, we have: $s = 3r$, $t = 2r$ [from equations (1) and (3)]; $x = (2/3)r$ [from equation (5) and the fact that $t = 2r$]; $v = (1/2)r$ [from equation (2)]; $w = (1/2)r + 2$ [from equations (2) and (4)].

So equation (6) can be written entirely in terms of r as: $r + 3r + 2r + (1/2)r + [(1/2)r + 2] + (2/3)r = 48$. Simplify this equation to obtain $(23/3)r = 46$; $r = 6$. The other values can be easily determined using the fact that $r = 6$.

A Job Not Done on Time: 3 days. Say $x =$ the part of the total job done by one dwarf in one day, so that $4x =$ the part of the job done by four dwarves in one day. Say $y =$ the number of days scheduled to complete the job.

The job was done in $(y + 2)$ days. All four dwarves worked one of those days, but, after that, only two dwarves worked. These two dwarves worked $(y + 2) - 1$ days, which simplifies to $y + 1$ days. If the four dwarves had worked until the end of the job the number of days scheduled to finish the job after one day of work would have been $y - 1$ days.

So in $y + 1$ days two dwarves did the amount of work that four dwarves would have done in $y - 1$ days. Hence: $2x(y + 1) = 4x(y - 1)$. Divide both sides by $2x$ to obtain: $y + 1 = 2(y - 1)$; $y = 3$.

A Lame Horse: 10 times. The knight walked two-thirds of the distance and rode one-third of it, so he walked twice as far as he rode. The walking portion took twenty times as long as the riding portion; so, if the walking and riding distances had been the same, the walking part would have taken ten times as long as the riding part. Thus, the knight rode ten times as fast as he walked.

A Round Table Arrangement: 1, Hob; 2, Rob; 3, Cob; 4, Bob; 5, Tob; 6, Lob.

Cob doesn't have seat #1 (clue 3). From clue 2, either Hob or Lob has that seat. Suppose it is Lob who has the #1 seat. Then Hob would have the #6 seat and Cob the #4 seat (clue 2). Rob would then have seat #3 (clue 3), and Bob would have the #5 seat (clue 1). By elimination, Tob would have seat #2, contradicting clue 4.

So Hob must have seat #1. Cob and Lob have seats #3 and #6 in one order or the other (clue 2). Suppose Lob has seat #3 and Cob seat #6. Then, by clue 1, Bob would have seat #5, while by clue 3, Rob would have seat #5. Thus, Cob has seat #3 and Lob seat #6. By clue 3, Rob has seat #2. Bob has seat #4 (clue 1). By elimination, Tob has seat #5.

A Wizard Competition: 48 butterflies. Say Bu = the number of butterflies. Say A, B, and C = the number of handsome men after the first, second, and third stages of the competition, respectively. Say X, Y, and Z = the number of beautiful princesses after the first, second, and third transformations, respectively.

The following equations follow from the information given:

(1) $A + X = Bu$; $B + Y = Bu$; $C + Z = Bu$

(2) $B = A + \frac{1}{3}X - \frac{2}{5}A = \frac{3}{5}A + \frac{1}{3}X$

(3) $Y = X + \frac{2}{5}A - \frac{1}{3}X = \frac{2}{3}X + \frac{2}{5}A$

(4) $B = Y$

(5) $C = B + (\frac{3}{4})(\frac{2}{5})A - (\frac{1}{2})(\frac{1}{3})X = B + \frac{3}{10}A - \frac{1}{6}X$

(6) $Z = Y + (\frac{1}{2})(\frac{1}{3})X - (\frac{3}{4})(\frac{2}{5})A = Y + \frac{1}{6}X - \frac{3}{10}A$

(7) $C = Z + 12$.

Since $B = Y$, the right hand sides of equations (2) and (3) can be made equal to obtain equation (8): $\frac{3}{5}A + \frac{1}{3}X = \frac{2}{3}X + \frac{2}{5}A$

Solving equation (8) for A gives $A = \frac{5}{3}X$. By substituting $\frac{5}{3}X$ for A into equation (2), we get $B = (\frac{3}{5})(\frac{5}{3})X + \frac{1}{3}X$; $B = \frac{4}{3}X$.

Next, equation (5) can be rewritten as equation (9): $C = \frac{4}{3}X + (\frac{3}{10})(\frac{5}{3})X - \frac{1}{6}X = \frac{5}{3}X$.

Similarly, by substituting $\frac{5}{3}X$ for A in equations (3) and (6), we can establish equation (10): $Z = X$.

Thus, from equation (7), $\frac{5}{3}X = X + 12$; $X = 18$. From equation (9), $C = 30$. From equation (10), $Z = 18$. From equation (1), $Bu = 30 + 18 = 48$.

Able and Gallant Soldiers: Able has 10 mounted soldiers and 14 foot soldiers. Gallant has 14 mounted soldiers and 12 foot soldiers.

Together, the two armies have fifty troops. Since Abel has two fewer troops in total than Gallant, Able has 24 troops, and Gallant has 26. Say Am and Gm = the number of mounted soldiers in each army and, Af and Gf = the number of foot soldiers in each. The information given leads to these four equations: (1) $Am + Af = 24$; (2) $Gm + Gf = 26$; (3) $Gm = Af$; (4) $Am = Gm - 4$.

Since Gm = Af (equation 3), we can substitute Af for Gm in equation (4) to obtain Am = Af – 4. Am + Af = 24 (equation 1); so when Af – 4 is substituted for Am in that equation, it becomes (Af – 4) + Af = 24. Hence, Af = 14. Thus, Am = 10. From equation (4), Gm = 14; so Gf = 12.

Apples for All: 16 apples, 8 boys. Let x = the original number of boys, so that 3x = the number of apples first picked. The final number of boys was x + 3 and the final number of apples was 3x + 1. Thus, (3x + 1) ÷ (x + 3) = 2. Simplified, this equation becomes 3x + 1 = 2x + 6. So x = 5 and 3x = 15. After three more boys appeared, there were eight boys to divide sixteen apples, so each boy got two apples.

At What Time Does the Wagon Driver Leave His Hut?: The driver leaves his hut at 3:50 P.M. and arrives at the dock at 5:05 P.M. In the 10 minutes from 4:05 to 4:15, he goes $\frac{1}{3}$ – $\frac{1}{5}$ = $\frac{2}{15}$ the distance. So in 5 minutes he goes $\frac{1}{15}$ the distance, and in 15 × 5 = 75 minutes he goes the full distance. At 4:15 he has made $\frac{1}{3}$ of the trip, which has taken $\frac{1}{3}$(75) = 25 minutes. So he leaves his hut at 3:50 and arrives at the dock 75 minutes later, at 5:05.

Building a Bridge: 15 hours. For each hour that Dobbit worked, $\frac{1}{30}$ of the project was completed. So after Dobbit had worked 5 hours alone, 5($\frac{1}{30}$), or $\frac{1}{6}$, of the job was completed, leaving $\frac{5}{6}$ of the job for Dobbit and Mobbit to do together. For each hour that Mobbit worked he did $\frac{1}{45}$ of the job. For each hour the two worked together the part of the job that was done was $\frac{1}{30}$ + $\frac{1}{45}$ = $\frac{5}{90}$, which reduces to $\frac{1}{18}$. So, together, the two would do the entire job in 18 hours. Thus, to do $\frac{5}{6}$ of it required $\frac{5}{6}$ × 18 = 15 hours.

Chased by a Glub: Since Phipos first ran half the time it took him to reach the fort, he ran for more than half the distance. So, when half the distance had been covered, Phipos was still running, but the glub had begun to walk. Therefore, the glub fell farther and farther behind Phipos. When Phipos began to walk, the glub was still walking, so the distance Phipos gained while he ran and the glub walked was maintained.

Contests of Skill: 6 days. Say E, M, and K = the number of elves, Minotaurs, and knights, respectively. From the clues: (1) $M = E - 1$; (2) $K + M = 9$; (3) $EK = 72/3 = 24$; $K > E$.

One solving method is to rewrite equation (1) as: (1a) $M - E = -1$.

Then subtract equation (1a) from equation (2) to obtain $K + E = 10$. Since in equation (3) $EK = 24$ and $K > E$, $K = 6$ and $E = 4$. So $M = 3$. The number of elf vs. Minotaur competitions was $4 \times 3 \times 3 = 36$; the number of human vs. Minotaur competitions was $6 \times 3 \times 3 = 54$. Then, $72 + 36 + 54 = 162$ competitions. Divide 162 by 27 to get 6 as the number of days necessary to complete the tournament.

Drawing Stones: 8 black, 4 white. Say n is the total number of stones, and b and w are the number of black and white stones, respectively, that Gopar put into the container. The probability of drawing a black stone on the first draw is $b/(b + w)$. After drawing out two black stones, the number of black stones remaining in the jar is $b - 2$. So the probability of drawing a third black is $(b - 2)/(b + w - 2)$.

From the statement of the puzzle, $(b - 2)/(b + w - 2) = (9/10)b/(b + w)$. Simplifying, this becomes: $10(b - 2)(b + w) =$

$9b(b + w - 2)$, or $10b^2 + 10bw - 20b - 20w = 9b^2 + 9bw - 18b$.

We know $w = n - b$, where n is between 7 and 13. Substituting this in for w, we get:

$10b^2 + 10b(n - b) - 20b - 20(n - b) = 9b^2 + 9b(n - b) - 18b$, or $10b^2 + 10bn - 10b^2 - 20b - 20n + 20b = 9b^2 + 9bn - 9b^2 - 18b$, which reduces to $10bn - 20n = 9bn - 18b$. So, $bn + 18b = 20n$, or $b = 20n/(n + 18)$.

The only value of n that makes b an integer for n between 7 and 13 is $n = 12$, where $b = 8$. This makes $w = 4$.

Feeding the Horses: Yes. Sixty-nine days of winter remain, and enough food is on hand to feed the two remaining horses for 72 days. This puzzle can be solved using algebra, but I like the following solution: When the four horse were sold, Distal had been feeding his original six horses for 6 days. Had he kept all six horses, he would have been able to feed them for another 24 days. But as he had only two horses—one-third as many—the food will last three times as long, or 72 days.

Going to the Fair: Clack rode 4 hours and rested 8 hours, while Bopp rode 8 hours and rested 4 hours.

Say b = the length of time that Bopp rode (and Clack rested) and c = the length of time that Bopp rested (and Clack rode). Then (1) $b + c = 12$.

If r = Bopp's rate in miles per hour, then Clack's rate was 2r. Using the formula $d = rt$ (distance equals rate multiplied by time) we have $d = rb$ and $d = 2rc$. (The distance was the same for each man, so d is the same in both equations.) Hence, $rb = 2rc$ and $b = 2c$. (In other words, since Bopp rode half as fast as Clack, he rode for twice as long.)

Substituting 2c for b in equation (1) gives $3c = 12$. So $c = 4$. Then, from equation (1), $b = 8$.

How Can Everyone Cross the River?: The half-elf rows the human across to Tok, leaving the dwarf and the elf at Ak. The half-elf leaves the human at Tok and returns to Ak alone. On his next trip the half-elf rows the dwarf across, leaving the elf alone at Ak. The half-elf leaves the dwarf in Tok and returns to Ak with the human. Leaving the human at Ak, the half-elf rows the elf across the river and leaves the elf with the dwarf at Tok. On the last trip the half-elf takes the human across again to Tok.

How Did the Archers Cross the River?: First, the two children rowed the boat across. One child remained on the river's far side while the second child rowed back to the archers. An archer then rowed across alone. This archer sent the boat back with the child who had remained on the far side of the river. The two children then rowed across again, and again one remained on the far side while the other rowed back. Next, a second archer rowed across, and sent the boat back with the child who had been left on the far side of the river. This process was repeated until all the archers were across.

How Early Was the Barge?: 20 minutes early. The cart driver traveled 24 minutes less than usual. If he had not met the rider, he would have needed half of those minutes, 12 minutes, to get to the dock at his usual time. But the rider had ridden for 8 minutes before he and the cart driver met, so the barge was $12 + 8 = 20$ minutes early.

How Far Apart Were the Dragons?: 5 miles. To solve this puzzle, the formula d = rt (distance = rate multiplied by time) is used. The dragons' rates are stated as miles per hour, so time must be expressed in terms of hours in order to obtain a meaningful answer. Five minutes = $\frac{1}{12}$ hour, so the distance Argothel walked in 5 minutes at a rate of 24 miles per hour was 2 miles ($24 \times \frac{1}{12}$), while the distance Bargothel walked was 3 miles ($36 \times \frac{1}{12}$). So the dragons were 5 miles apart 5 minutes before they met.

How Far from Castleton to Devil's Peak?: Castleton and Devil's Peak are 10 miles apart. The rider who started his journey at Devil's Peak rode 1½ times as fast as the rider who started his trip at Castleton.

The diagram below shows that when the horsemen met for the first time, they had, together, traveled a distance equal to the distance between Castleton and Devil's Peak. When they met for the second time, they had traveled, together, three times the distance between the two towns.

Both riders traveled at a constant speed, so when they met the second time, each had ridden three times as far

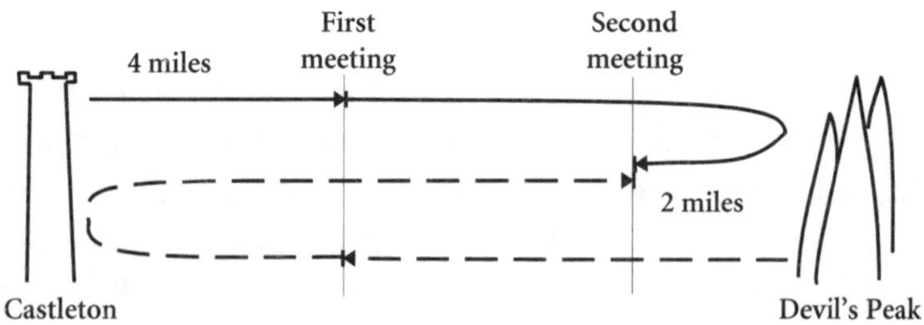

as he had when the two met the first time. The rider who began at Castleton had, thus, traveled $3 \times 4 = 12$ miles. The distance of 12 miles is 2 miles more than the distance between the towns; so, the towns are 10 miles apart.

As for the two riders' relative speed, during the time that the rider who began the trip at Castleton rode 4 miles, the other rider rode 6 miles. So the other rider rode $1\frac{1}{2}$ times as fast.

How Fast Does Lucien's Horse Need to Run?: Two-fifths as fast as Mnemo's horse. If r = the rate of Mnemo's horse in miles per hour, the rate of Wister's horse is $(\frac{1}{4})r$. The time needed for Mnemo's and Wister's horses to run 2 miles is the sum of both their times for a one-mile run.

Using the formula time = distance divided by rate, it will take Mnemo's horse $\frac{1}{r}$ hours to run one mile and Wister's horse $1 \div (\frac{1}{4})r$, or $\frac{4}{r}$ hours. Their combined time is $\frac{1}{r} + \frac{4}{r}$ hours, which is $\frac{5}{r}$ hours. Say the rate of Lucien's horse in miles per hour is x. Then its time to run the 2-mile race is $\frac{2}{x}$ hours. To finish the race at the same time as Wister's and Mnemo's horses, we must have $\frac{2}{x} = \frac{5}{r}$. Solving for x gives the answer $x = (\frac{2}{5})r$.

For example, say Mnemo's horse runs at a rate of 20 mph, and Wister's horse runs at a rate of 5 mph. Then Mnemo's horse needs $\frac{1}{20}$ hour (3 minutes) to run a mile, while Wister's horse will need $\frac{1}{5}$ hour (12 minutes) to run the same distance. Together, they will finish the 2-mile course in 15 minutes. If Lucien's horse runs at a rate of $\frac{2}{5}$ Mnemo's rate, which is 8 mph, he will also finish in 15 minutes, since $2 \div 8 = \frac{1}{4}$ hour (15 minutes).

How Long Are the Glubs?: The first, third, and fifth were each 40 feet long; the second and fourth were each 30 feet long. The distances between the glubs account for 20 of the 200 feet. So the five glubs together have a length of 180 feet. Say x stands for the length of each of the first, third, and fifth glubs. If the second and fourth glubs are each 10 feet longer than the others, then each is x + 10 feet long, and:

$3x + 2(x + 10) = 180; x = 32$

But 32 is not a multiple of 10, so the second and fourth glubs must each be 10 feet shorter than the others. Hence, each is x – 10 feet long, and:

$3x + 2(x - 10) = 180; x = 40.$

Since each of the first, third, and fifth glubs is 40 feet, the others are each 30 feet long.

How Many Cakes?: 24 cakes; each being took 6. Heartnik took ¼ the original number, leaving ¾ of that number. Scowler took ⅓ of that ¾, or ¼ of the original number. So after Heartnik and Scowler had helped themselves, ½ the original number of cakes remained on the plate. Goodin took ½ of that half, so he, as well as Heartnik and Scowler, took ¼ of the original number. Thus, after Goodin had helped himself, ¼ the original number remained. So Loglob, the last being to take the cakes, took ¼ the original number. As he took six cakes, there were 24 cakes in the beginning, and each being took six.

How Many Dwarves?: 36. The amphitheater will hold 120 humans, so 90 humans use 90/120, or three-fourths, of its capacity. This leaves one-fourth of its capacity for dwarves. Since 144 dwarves will fill the amphitheater, the number of dwarves that will fill one-fourth of it is (¼)144, or 36.

How Many Eggs?: 33. The number of eggs broken is an odd number—since, when divided by 2, there is a remainder of 1. The number is a multiple of 3, and when divided by 5 there is a remainder of 3.

So, for some positive integer k, $5k + 3$ = the number of eggs. Since the total number of eggs is an odd number, k must be an even number. If k is 8 or larger, the number of eggs is 43 or greater, since $(8 \times 5) + 3 = 43$. The given information states that the number of eggs is less than 40, so k is 2, 4, or 6.

If $k = 2$, then $5k + 3 = 13$.
If $k = 4$, then $5k + 3 = 23$.
If $k = 6$, then $5k + 3 = 33$.

Since the number of eggs broken is a multiple of 3, the answer is 33.

How Many Handshakes?: 105. The first knight shakes the hand of fourteen other knights. The second, having already shaken hands with the first, shakes the hand of thirteen others; the third, having shaken hands with the first and second, shakes hands with twelve others, and so forth. So we have the answer: $14 + 13 + 12 + 11 + 10 + 9 + 8 + 7 + 6 + 5 + 4 + 3 + 2 + 1 = 105$.

How Many Horses?: 15 horses. Let h = the number of horses and m = the number of men. Then (1) $h + m = 45$; so, $m = 45 - h$; and (2) $4h + 2m = 120$.

Substituting $45 - h$ for m in equation (2) gives: $4h + 2(45 - h) = 120$; $h = 15$. So Mistar has 15 horses (and 30 men).

How Many Lucs?: Jarblek, 16; Garion, 12; Belgar 8; Polgar, 4. Say x = the number of lucs Jarblek earned. Since the four men worked at the same rate, Belgar earned $(\frac{1}{2})x$, Polgar earned $(\frac{1}{4})x$, and Garion earned $(\frac{3}{4})x$. So: $x + (\frac{1}{2})x + (\frac{1}{4})x + (\frac{3}{4})x = 40$.

Adding, with 4 as the common denominator, gives: $(\frac{10}{4})x = 40$; $x = 16$.

The rest follows easily by substitution.

How Many Schlockels?: Altus has 32 schlockels. From clue 1, if the number of schlockels Altus has is a multiple of 5, the number is 5, 10, or 15. However, from clue 2, the number is not 5, 10, or 15, because none of these is a multiple of 8 and none is between 20 and 29. So the number of schlockels Altus has is not a multiple of 5. From this we know that it is also not a multiple of 10—since any number that is a multiple of 10 is a multiple of 5. Hence, from clue 3, Altus has 31, 32, 33, 34, 36, 37, 38, or 39 schlockels. The number cannot be "not a multiple of 8," so it is a multiple of 8. Thus, Altus has 32 schlockels.

How Many Winged Cats?: 2 winged cats. Say c is the number of winged cats that Gareth has. His reply provides the equation $(\frac{2}{3})c + \frac{2}{3} = c$; $\frac{2}{3} = (\frac{1}{3})c$. Multiply both sides of the equation by 3 to obtain the answer $c = 2$.

How Much Did Alaranthus Weigh?: 3000 pounds. Alaranthus's weight was such that 1000 pounds was equal to one-third of his weight. So his total weight in pounds was 3×1000.

How Much Does the Fully Grown Alaranthus Weigh?: 4000 lbs. Say a = Alaranthus's weight and m = Montal's weight. Then (1) $a = m + (\frac{1}{5})a$ and (2) $a = m + 800$.

Substituting $m + 800$ for a in equation (1) gives: $m + 800 = m + (\frac{1}{5})a$; so $800 = (\frac{1}{5})a$, and $a = 4000$.

How Much for a Knife, Sword, or Arrow?: 2 coins for a knife; 3 coins for an arrow; 7 coins for a sword. Say x, y, and z are the costs in coins of one arrow, one knife, and one sword, in that order. Then, (1) $36 = y + z + 9x$; (2) $2z = y + 4x$, or $y = 2z - 4x$.

Substituting $2z - 4x$ for y in equation (1) leads to $36 = 2z - 4x + z + 9x$. This simplifies to equation (3): $x = (36 - 3z) \div 5$.

Normally, solving one equation containing two variables will not produce a unique answer. However, in this case we know that x, y, and z are all positive integers, and, from equation (3), the quantity $36 - 3z$ is evenly divisible by 5 (since x is a positive integer).

Trial and error quickly shows that the positive integral values of z that make $36 - 3z$ evenly divisible by 5, and x a positive integer, are $z = 2$ and $z = 7$. If $z = 2$, then $x = 6$; but if $x = 6$, then from equation (1) or (2) y would be -20, which is not possible. So $z = 7$; hence $x = 3$, $y = 2$.

How Old Is the Wizard Alchemerion?: 360 years old. Say A = Alchemerion's age; S = his son's age; F = his father's age. Then: (1) $A = 3S$; (2) $F = 2A + 40$; (3) $A + S + F = 1240$.

From equation (1) $S = (\frac{1}{3})A$. Substituting $2A + 40$ for F and $(\frac{1}{3})A$ for S in equation (3) gives us: $A + (\frac{1}{3})A + (2A + 40) = 1240$; $A = 360$.

How Old Is Mongo's Son?: 10 years old. If a is the son's age and x is the number which has an integral square root, then: (1) $a + 6 = x$; (2) $a - 6 = \sqrt{x}$.

Trial and error works well here. From equation (2), the age, a, is larger than 6, since the square root of x is a positive number. So, substitute the numbers 7, 8, 9, etc. for a in equation (1) until you find a number for x that is a perfect square. Find its square root and see if it checks in equation (2). The first value for a that is larger than 6 and which makes $a + 6$ a perfect square is 10. Therefore, $x = 16$. These values check when substituted into equation (2).

{An algebraic solution requires solving the quadratic equation $(a - 6)^2 = a + 6$.}

Human vs. Minotaur: The human warrior's clockwise counting should begin with the warrior third from the Minotaur moving clockwise. He numbers the positions 1 to 7 as in the arrangement below, with M, the Minotaur's position, labeled "1." He would then start counting seven places beginning with "4." This lands him on "3." He defeats "3," then, beginning with "4," he counts clockwise another seven places. This brings him around the circle to "4," with whom he fights the second battle. Continuing in this manner, the order in which the warrior will fight his foes is 3, 4, 6, 2, 5, 7, 1.

M
1
7 2
6 3
5 4

Inspecting the Troops: 59 seconds. The answer is not 58 seconds (twice 29 seconds). Refer to the distance between the first and second troop, or second and third troop, etc., one *segment.* There are twenty-nine segments between the first and thirtieth man. The time required was 29 seconds to cover this distance, so the officer rides at the rate of 1 second per segment. There are fifty-nine segments in total, so the total time required will be 59 seconds.

Jousting Tournament Number: Sir Bale's number was 5. Four numbers are less than 5: 1, 2, 3, and 4. Six numbers are greater than 5: 6, 7, 8, 9, 10, 11. The product of 4 and 6 is 24. The answer would be the same if Sir Bale's number were 7; then there would be six numbers less than 7 and four numbers greater than 7.

Measuring Two Gallons of Cider: Fill the 3-gallon container with cider and empty it into the 4-gallon container. Fill the 3-gallon container a second time and pour it into the 4-gallon container. When the 4-gallon container is full, 2 gallons remain in the 3-gallon container.

Meeting the Stone Cutter: 5:50 A.M. The cart would normally have arrived at the ferry landing at 6:00 A.M. But on the day in question, the cart driver delivered the stone worker to the quarry 20 minutes early, so the driver spent 20 minutes less time traveling than he usually did: 10 minutes going toward the ferry and 10 minutes going to the quarry. Subtract 10 minutes from 6:00 A.M. to get 5:50 A.M. as the time the cart met the worker.

If you find this difficult to follow, let's say that normally the driver leaves the quarry at 5:00 A.M. in order to arrive at the ferry at 6:00 A.M. and deliver the worker to the quarry at 7:00 A.M. In other words the cart driver drives one hour each way on a normal day. But on the day in question, the cart driver delivered the worker to the quarry at 6:40. The cart driver left for the ferry at his usual time, 5:00 A.M., and drove until 6:40, so he drove for 100 minutes, 50 minutes in one direction and 50 in another. Normally he drives 60 minutes to reach the ferry landing. Thus, he drove for 10 fewer minutes toward the ferry. So he must have picked up the worker at 5:50 A.M. rather than 6:00 A.M.

Minotaur Fighters: Logi had 78, Magnus had 42, Nepo had 24. The best way to solve this problem is by making a table that begins at the end of the lending:

Logi		Magnus		Nepo	
48	+	48	+	48	= 144
↓		↓		↓	
24	+	24	+	96	= 144
↓		↓		↓	
12	+	84	+	48	= 144
↓		↓		↓	
78	+	42	+	24	= 144

More Jousters: 5 knights originally; 4 unknown knights. Say x = the number of competitors before the change effected by the entrance of y more knights. The number of jousts with x competitors is $[x(x - 1)] \div 2$. (For example, suppose there are four competitors. Each one of the four jousts against each of the other three. Now $4 \times 3 = 12$, but when A fights B, B is also fighting A, so 12 must be divided by 2 to get 6 as the answer to the questions of how many jousts there are with four entrants: $[4(4 - 1)] \div 2 = 6$).

With y knights added to the original x knights, the number of jousts becomes $(x + y)(x + y - 1) \div 2$. So, $[(x + y)(x + y - 1) \div 2] - [x(x - 1) \div 2] = 26$. Simplified, this becomes equation (1): $y(2x + y - 1) = 52$.

Since both x and y (the number of original and new competitors, respectively) are positive integers, $2x + y - 1$ is also a positive integer. Thus, from equation (1), y is a factor of 52. It is 1, 2, 4, 13, 26, or 52. If y = 1, then the puzzle description would not have referred to "some" unknown knights. If y = 13, 26, or 52, x becomes a negative number. So y is either 2 or 4. If y = 2, then x = 12½—not a positive integer. So y = 4. Substituting 4 for y in equation (1) gives x = 5.

More Mug Clinking: Algar's archenemy clinked mugs with Daring, Elveron, and Grace. Daring clinked his mug with four others. Remember that seven different answers were given to the Algar's question. No knight clinked mugs with his archenemy or with himself, so the maximum number of mugs that any one knight could clink was six. So the number of mugs a particular knight clinked was from zero to six.

In the diagram below, the eight knights (with Algar included) are indicated by the number of mugs they clinked. For example, "6" is the knight who clinked mugs with six other knights. A connecting line indicates that two knights had clinked mugs.

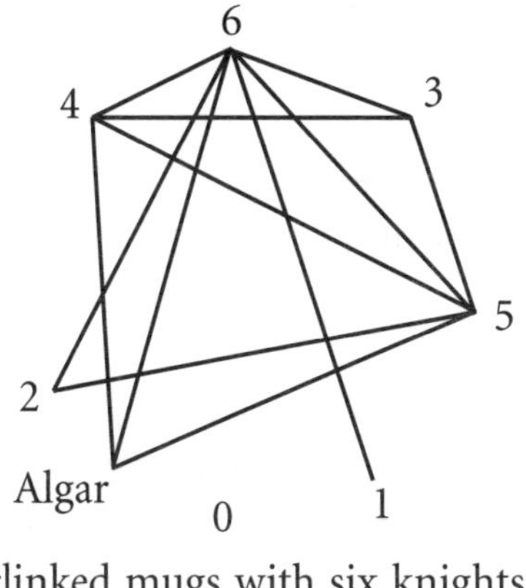

Since "6" clinked mugs with six knights, he is connected to all the others knights except "0."

"1" clinked mugs with one person, so he must be connected only to "6."

"5" is connected to all but "0" and "1."

The joining done so far has linked "2" with both "6" and with "5," so "2" clinked with no other mugs.

In addition to being joined to "5" and "6," "4" must, by elimination, be connected to 3 and to Algar. So "3" is connected to "6," "5," and "4," and not to "0," "1," or "2."

Thus, the diagram shows that "6" and "0" are archenemies. Hence, "5" and "1" are archenemies. So "4" and "2" are another pair of archenemies, and Algar's archenemy is "3." The diagram also shows that Algar and his archenemy clinked precisely the same mugs—numbers "4," "5," and "6." We are told that the only mug Algar clinked with that belonged to a knight of Dracon was Daring, whose archenemy was Cautious from King Wincher's court.

Added together, 0, 1, 2, 3, 4, 5, and 6 total 21. So, from what we are told, the total number of mug clinkings done by the three King Wincher knights: Cautious, Elveron, and Grace, was 13 and the total number of mug clinkings done by the four knights of King Dracon was 8.

Thus, "6," "5," and "2" or "6," "4," and "3" are the other knights of King Wincher's court. Since "3" is Algar's archenemy, he is not a knight of King Wincher; so "6," "5," and "2" are the other King Wincher knights.

Algar, who clinked mugs with "4," "5," and "6," clinked mugs with Daring of King Dracon's court; so Daring is "4."

We know "4" and "2" are archenemies and have been told that Daring's archenemy is Cautious. So "2" is Cautious, and "5" and "6" are Elveron and Grace in one order or the other.

Mugs Clinking: 15. Each of the fifteen knights clinked mugs with two other knights. 15 × 2 = 30. But 30 must be divided by 2 to get the correct answer of 15, because, if knight A clinked mugs with knight B, then knight B also clinked mugs with knight A.

Perforce Arrives Too Late: The march had been planned to take 15 days, but it took 20 days. Say x = the number of days the march was originally planned to take. It actually took x + ($\frac{1}{3}$)x days. The extra ($\frac{1}{3}$)x days consisted of half a day caused by rain, ($\frac{1}{10}$)x day caused by the bridge washout, and three days caused by the fatigue of the army. So $\frac{1}{2}$ + ($\frac{1}{10}$)x + 3 = ($\frac{1}{3}$)x. Simplifying, $\frac{1}{2}$ + 3 = ($\frac{1}{3}$)x – ($\frac{1}{10}$)x; x = 15. The trip took one-third longer than the 15 days planned, so it took 20 days.

Rings for the Princesses: King Firnal can bestow two princesses with rings and then give the box with the remaining ring to the third princess.

Sharing a Job: Abelard 15 days; Brendan, 30 days; Cullen could never do the job.

Say a is the part of the job done by Abelard in one day; b is the part of the job done by Brendan in one day; c is the part of the job done by Cullen in one day. Since Abelard and Brendan require 10 days working together to complete the job, they do $\frac{1}{10}$ of the job per day, hence: (1) a + b = $\frac{1}{10}$. Similarly, (2) a + c = $\frac{1}{15}$; (3) b + c = $\frac{1}{30}$.

Subtract equation (2) from equation (1) to obtain (4) b – c = $\frac{1}{30}$. Add equations (3) and (4) to obtain 2b = $\frac{1}{15}$, b = $\frac{1}{30}$. Substitute $\frac{1}{30}$ for b in equation (1) to obtain a = $\frac{1}{15}$. Now, substitute $\frac{1}{15}$ for a in equation (2) to obtain c = 0. So Cullen does no work and could never get the job done; Abelard could do it alone in 15 days and Brendan could do it alone in 30 days.

Slaying Glubs: Grabus, 14; Hylar, 10. This puzzle is easily solved using trial and error. For an algebraic solution, say g = the number of glubs Grabus has slain and h = the number Hylar has slain. Then (1) g + h = 24 and (2) g = h + 4.

Substituting $h + 4$ for g in equation (1) gives $(h + 4) + h = 24$; $2h + 4 = 24$; $2h = 20$; $h = 10$. So $g = 14$.

Strange Rabbits: 5 jumps. One way for the rabbit to reach the carrot is to take four 5-foot jumps forward (putting it 7 feet in front of the carrot), then one 7-foot jump backward. Of course, there are several possible variations of this, all really the same, just in a different order (for example, the rabbit could jump 5 feet forward, then 7 feet backward, then take three 5-foot jumps forward).

To solve this problem algebraically, let x and y be the number of 5-feet and 7-feet jumps, respectively, with a positive x or y referring to a forward jump and a negative x or y referring to a backward jump. So, an algebraic solution amounts to finding integral values for x and y that satisfy $5x + 7y = 13$. The solution that requires the fewest jumps is $x = 4$, $y = -1$.

Switching Allegiance: Lavar has seven and Malcavar has eleven. Before Colin made his change from Malcavar's troop to Lavar's, Malcavar had two more soldiers than Lavar (since Colin's switch made the number of soldiers in the two troops equal). When Colin switched back to Malcavar's troop, once again Malcavar had two more soldiers than Lavar. Then Draal moved from Lavar's troop to Malcavar's, decreasing Lavar's number of soldiers by one and increasing Malcavar's by one; so, Lavar then had four fewer soldiers than Malcavar.

Together the two men had more than ten and less than thirty soldiers. The only numbers that fit the requirements are seven and eleven.

At first, Lavar had eight and Malcavar had ten soldiers. Once Colin switched from Malcavar to Lavar, each had nine. When Colin switched back, Lavar had eight soldiers again and Malcavar had ten soldiers. Once Draal moved from Lavar's troop to Malcavar's, Lavar had seven soldiers and Malcavar had eleven.

The Cost of Cider: 7 stickels. Say x = the cost of the liter container. Then x + 4 is the cost of the cider. So x + (x + 4) = 10. When simplified, this equation becomes 2x = 6; x = 3. So the container costs 3 stickels; hence, the cider costs 7 stickels.

The Daughters of Alexis: The girls are 1, 2, and 8 years old. Blythe could not answer the question using the first clue because ten different combinations of three numbers add to 11 (1, 1, 9; 1, 2, 8; 1, 3, 7; 1, 4, 6; 1, 5, 5; 2, 2, 7; 2, 3, 6; 2, 4, 5; 3, 3, 5; 3, 4, 4).

Since the second clue did not provide Blythe with the answer, it must be the case that at least two combinations from the list above have a product which is either 16 years more or 16 years less than Blythe's age. The products derived from the list above are, in the order of the list, 9, 16, 21, 24, 25, 28, 36, 40, 45, 48. Since no two of these products are the same, it must be true that one product is 16 more and the other is 16 less than Blythe's age. Thus, the difference between these two products is 32.

Comparing each product with each of the others, we find two products that meet this requirement: 16 and 48, the products of 1, 2, and 8, and 3, 4, and 4, respectively (thus, Blythe is 32 years old). Blythe could not determine the children's ages even with this second clue. However, Alexis's third statement revealed that there was a daughter whose age was greater than the others, so the 3, 4, 4 combination was ruled out.

The Farmer and the Hobgoblin: 15 coins. To solve the puzzle start at the end. Since the farmer gave the hobgoblin his "last sixteen coins," sixteen is the number of coins the farmer had after the final doubling; so, he had eight coins when he crossed the field for the last time. Add to this the sixteen he gave the hobgoblin after the third crossing, and the result is twenty-four, which is twice the number of coins he had before the third crossing.

So the farmer had twelve coins when he started the third crossing. Adding to twelve coins the sixteen he gave the hobgoblin after the second crossing gives twenty-eight coins, which is twice the number he had before the second crossing.

So he started the second crossing with fourteen coins. Add to fourteen the sixteen coins he gave the hobgoblin after the first crossing, and the result is thirty, which is twice the number of coins he started with. Thus, the farmer began the first crossing with fifteen coins.

The Fools' Show: 2 knights; 10 children; 18 other adults. Since knights paid 25 stickels per ticket and other adults paid 50 stickels per ticket, the total amount that these two groups paid is evenly divisible by 25. Since 1000 (total receipts) is also divisible by 25, the difference between 1000 and the total amount paid in by knights and "other" adults is also evenly divisible by 25. In other words, the amount that children paid for tickets is evenly divisible by 25. If x is the number of children who paid, that amount is 5x; so 5x = 25k, with k being a positive integer. Thus x = 5k, showing that the number of children who paid is evenly divisible by 5.

Since the number of other adults (non-knights) who paid was greater than the number of children who paid, and the total number of ticket buyers was thirty, fewer than fifteen children paid for admission. Thus, either five or ten children's tickets were sold. If five were sold, then nine or fewer other adults bought tickets, and four or fewer knights bought tickets. That would mean the total number of ticket buyers was at most eighteen. So ten children's tickets were sold.

The total receipts from children's tickets were $10 \times 5 = 50$ stickels. Together, knights and other adults paid in 950 stickels, and the number of knights and other adults totaled 20. If y = the number of knights and z the number of other adults: (1) $y + z = 20$; (2) $25y + 50z = 950$.

From equation (1), $y = 20 - z$. Substituting $20 - z$ for y in equation (2) leads to the equation: $25(20 - z) + 50z = 950$; $z = 18$. So $y = 20 - 18 = 2$.

The Island of Odds: $\frac{5}{9}$ or 55.6%. Let T, L, and N stand for truth teller, liar, and normal, respectively. The first person is equally likely to be any one of the three types. The same is true of the second, since the type of native met on the first day is possibly the same type met on the second. There are nine combinations for the 2 days: LN (liar followed by normal), LL (liar followed by liar), LT (liar followed by truth teller), and, using the same symbols, NN, NL, NT, TN, TL, and TT. In five of these nine cases, one or both are normals.

The Jewel Chest: 17. Say x = the number of pins in the chest. Then the number of rings is $(\frac{5}{2})x$. The number of pairs of earrings is $(\frac{5}{2})x - 4$ and the number of individual earrings is twice that: $2[(\frac{5}{2})x - 4]$, which, simplified, becomes $5x - 8$. The total number of pieces of jewelry is 26; so: $x + (\frac{5}{2})x + (5x - 8) = 26$; $x = 4$.

Thus, Amarina has ten rings and six pairs of earrings—three pairs by one artisan and three pairs by another.

Therefore, there are six individual earrings of one kind and six of another.

It is possible that all the pins and rings will be removed before an earring is removed, so that the first earring removed is the fifteenth item removed. Then the sixteenth item might be a non-matching earring. The seventeenth item removed has to match either the fifteenth or sixteenth item removed.

The Messenger: It is not possible. The first 16 miles of the trip took 2 hours of running. However, the entire trip had to be run in 2 hours if the average rate for the entire trip was to be 12 miles per hour. So the task is impossible.

The Price of Candy: 19 pieces of candy: 9 bonbons at 9 shukels each, 6 sweetles at 6 shukels each, 4 chocos at 4 shukels each. Say x is the number of bonbons purchased. Say y and z are, in one or the other order, the number of sweetles and chocos purchased. The bonbons cost x shukels each; either the sweetles or chocos cost y shukels each while the other costs z shukels each.

The total number of pieces purchased is $x + y + z$, so at an average cost of 7 shukels each, the total cost is $7(x + y + z)$. The total cost is also $x^2 + y^2 + z^2$ because Gwendar bought as many pieces of each candy as its price in shukels. So, $7(x + y + z) = x^2 + y^2 + z^2$.

From the above, we get equation (1): $7x + 7y + 7z = x^2 + y^2 + z^2$.

Since bonbons cost more per piece than the other types of candy, and since the average cost Gwendar paid per piece was 7 shukels, it follows that x, the cost of one bonbon, is larger than 7.

If $x = 8$, either y or $z = 5$ (since either a piece of sweetle or a piece of choco costs 3 shukels less than a piece of bonbon). Assume that $y = 5$. By substituting 8 for x and 5 for y in equation (1), we have: $56 + 35 + 7z = 64 + 25 + z^2$; $91 + 7z = 89 + z^2$; $2 = z^2 - 7z$, for $z = 1, 2, 3, 4, 6$, or 7.

None of these values of z make the equation true, so x cannot be 8.

Now try $x = 9$ and assume $y = 6$. Then, $63 + 42 + 7z = 117 + z^2$; $0 = z^2 - 7z + 12$.

Factoring the polynomial above gives us $0 = (z - 3)(z - 4)$; thus, $z = 3$ or $z = 4$. All three kinds of candy cost more than 3 shukels each, so $z = 4$ if $x = 9$ and $y = 4$.

The solution $x = 9$, $y = 6$, $z = 4$ is the only possible solution, because if $x \geq 10$, then $y \geq 7$, $z \geq 4$, so the total cost of $x^2 + y^2 + z^2 \geq 100 + 49 + 16$, which is 165, more than Gwendar spent.

To check our answers, $(9 \times 9) + (6 \times 6) + (4 \times 4) = 81 + 36 + 16 = 133$ shukels. 19 pieces of candy at 7 shukels each = 133 shukels.

The Shopping Trip: Mantle, 25 stickels; robe, 40 stickels; chest, 15 stickels.

Say m, r, and c are the costs in stickels for the mantle, robe, and chest, respectively. (The cost of the bracelet Marcella wanted is also m.) Then: (1) $r + c + m = 80$; (2) $r + m = 5c - 10$ (here m stands for the cost of the bracelet); (3) $r = c + m$.

Rewrite equation (1) as $r + (c + m) = 80$. Substitute $c + m$ with r from equation (3) to obtain the equation: $2r = 80$. So $r = 40$. Rewrite equation (1) as $(r + m) + c = 80$; then substitute $r + m$ with $5c - 10$ from equation (2) to obtain

$5c - 10 + c = 80$; $c = 15$. Thus, by substitution of the values of r and c into any of the original equations, $m = 25$.

The Sons of Blythe: They are 2, 3, and 6 years old. Since all three are younger than 10 years of age, and the product of the ages of the two youngest equals the age of the oldest, the ages are 2, 2, 4; or 2, 3, 6; or 2, 4, 8; or 3, 3, 9. Of these, only 2, 3, and 6 add up to a number that is a prime number.

The Wizards' Dragon Spells: Sagareth, 4 years; Thaumater, 16 years; Malefano, 64 years. Say M, S, and T represent the number of years the spells will last as cast by Malefano, Sagareth, and Thaumater, respectively. The information given leads to these equations:

(1) $S = \sqrt{T}$

(2) $S = \frac{1}{2}\sqrt{M}$

(3) $T = 2\sqrt{M}$

From equations (1) and (2), $\sqrt{T} = \frac{1}{2}\sqrt{M}$. Square both sides to get $T = \frac{1}{4}M$. Since $T = 2\sqrt{M}$ in equation (3), $\frac{1}{4}M = 2\sqrt{M}$. Squaring both sides, $\frac{1}{16}M^2 = 4M$. Multiply both sides by 16 to obtain $M^2 = 64M$. Thus, $M^2 - 64M = 0$. Factoring, $M(M - 64) = 0$. From this, $M = 0$ or $M = 64$. From the information given, $M = 64$. Thus, $S = 4$ from equation (2), and $T = 16$ from equation (1).

To the King's Castle: 8 miles. Alf's answer to Beryl's first question was that the distance from their cottage to the point where she posed the question was three times the distance in miles from that point to the inn. Let $x =$ the distance in miles from that point to the inn.

Alf's answer to Beryl's second question was that the distance to the castle from the point where she asked the second question was three times as far as the distance they had gone since leaving the inn. Let y = the distance in miles they had gone since leaving the inn. The distance between the points on the trip where the two questions were asked was 2 miles. The diagram below clarifies the puzzle.

Since x + y = 2, it follows that 3(x + y) = 6. But 3 (x + y) is the same as 3x + 3y, distances that are marked in the diagram. The total distance, therefore, is (x + y) + (3x + 3y) miles, which is 2 + 6 = 8 miles.

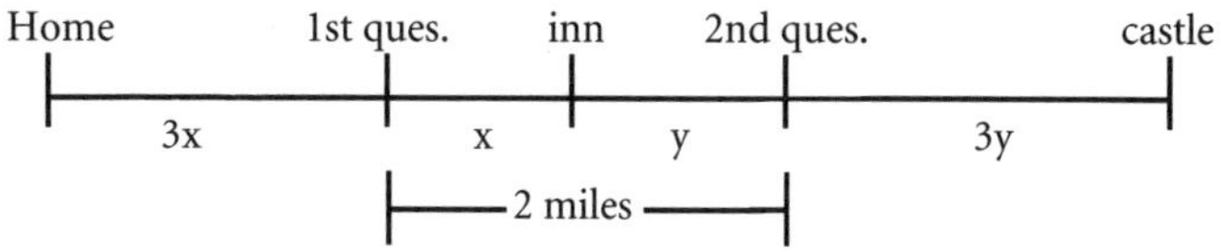

Tricky Squares: Dob is right. It cannot be done. Number the dots as shown in the diagram below:

o	o	o
1	2	3
o	o	o
4	5	6
o	o	o
7	8	9

Color dot number 5 white. If the diagonally opposite dots in the same square are to be different colors, then dot number 1 must be black (because dots 1, 2, 4, and 5 make a square) and dot 9 must be black (because dots 5, 6, 8, and 9 form a square). But dots 1, 3, 7, and 9 also form a square; so, the fact that 1 and 9 are the same color is a contradiction. A similar contradiction follows if dot number 5 is colored black.

Trinkets to Share: 12 trinkets, 5 girls. Let t = the number of trinkets and g = the number of girls. Then g is a factor of t (i.e., g goes into t without a remainder). Let x = the number of trinkets each girl received with the first distribution of t trinkets. So, (1): $gx = t$. If the t trinkets had been divided by families rather than by individuals, each share would have increased to x + 5 trinkets, and the number of recipients would have decreased to g – 2. Thus, we obtain (2): $(g - 2)(x + 5) = t$. When Felice's x trinkets were divided among the other girls, each girl received additional trinkets, but the division was still equal. So t, the number of trinkets, is evenly divisible by g – 1, as well as by g and by g – 2.

Since, by (1), $gx = t$, and, by (2), $(g - 2)(x + 5) = t$, it follows that $gx = (g - 2)(x + 5)$. Multiplying on the right gives: $gx = gx + 5g - 2x - 10$. From this, $5g = 2x + 10$; so (3): $g = (2/5)x + 2$.

Since $(2/5)x + 2$ is a positive integer, so also is $(2/5)x$. Thus, x is evenly divisible by 5, and g is greater than or equal to 4.

Make a table with x in the first column, g in the second, t in the third, g – 1 in the fourth, and g – 2 in the fifth. We are looking for a value for t that is evenly divisible by g, g – 1, and g – 2, with g less than 10. The table shows that there were 6 girls and 60 trinkets originally, so after Felice gave up her share, each of the 5 girls got 12 trinkets.

x	g [from (3)]	t	g – 2	g – 1
5	4	20	2	3
10	6	60	4	5
15	8	120	6	7
20	10 (too large a number)			

Two Riders: 58 miles. In an hour the first rider traveled 30 miles and the other 28 miles, so they were 58 miles apart one hour before they met each other.

Weighing a Pound of Flour: She put the one-pound weight on one pan of the scale and balanced it with lead pellets placed on the other pan. She then removed the one-pound weight and replaced it with flour until it balanced the lead pellets.

What Is the Human Population of South Pymm?: 389. The population of North Pymm is smaller than 500 and is a number that has 3, 4, 5, and 7 as integral factors (i.e., 3, 4, 5, and 7 divide into the number without a remainder). Since 3, 4, 5, and 7 are relatively prime, the smallest possible number is the product of 3, 4, 5, and 7, which is 420. The next largest possible number would be 840, which is too large. So the population of North Pymm is 420 and the population of South Pymm is $(\frac{1}{3})420 + (\frac{1}{4})420 + (\frac{1}{5})420 + (\frac{1}{7})420 = 140 + 105 + 84 + 60 = 389$.

Which Coin Is Lighter?: One way to accomplish the task is as follows. Peppi first divides the fifty coins into three groups: two of seventeen each and one of sixteen.

First Weighing: Peppi uses the scale to compare the two groups of seventeen. He places one group on one side of the scale and the other group on the other side. If one side is lighter than the other, the light coin has been identified as being in one particular group of seventeen coins. If the two sides balance, the light coin is in the group of sixteen that was not weighed.

Next, Peppi divides the group of sixteen coins, or the group of seventeen coins, whichever has been determined to contain the lighter coin, into three groups. If there are sixteen coins, he divides them into two groups of five coins each and one group of six coins. If there are seventeen coins, he divides them into two groups of six coins each and one group of five coins.

Second Weighing: Following the method outlined above Peppi identifies which one of the three groups contains the lighter coin. In other words, he balances the two groups of five coins against each other or the two groups of six coins against each other, depending on the result of the first weighing.

The second weighing identifies a group of either five or six coins that contains the lighter coin. If the light coin is in the former group, Peppi divides the five coins into two groups of two each, leaving one coin by itself. If it is the latter group, he divides the six coins into three groups of two each.

Third Weighing: Peppi compares two groups that have two coins each. If they balance on the scale, and only one coin remains in the third group, then the coin in the third group is the light one. If they don't balance, then the light coin is in the lighter group of two coins. If they do balance, the remaining group of two coins contains the lighter coin.

Fourth Weighing: If necessary, Peppi compares two single coins against each other to identify the light coin.

Who Rode Faster?: Sir Chaste rode faster. Suppose Pure rode x hours and Chaste rode y hours. So Pure rested $(1/4)y$ hours and Chaste rested $(1/3)x$ hours. In terms of time spent in making the trip, we have: $x + (1/4)y = y + (1/3)x$. Solving for x in terms of y gives $x = (9/8)y$, so x is larger than y. This means that Pure rode longer than Chaste to cover the same distance, so Chaste rode faster.

Did the Dragon Catch Pryor?: No, Pryor made it to the sea cave with less than four seconds to spare. If Wivere were to catch up with Pryor, he had to cover 7 miles before Pryor covered 2 miles. Pryor ran at a constant rate of 20 miles per hour, so he covered 2 miles in 6 minutes. As for Wivere, he ran the first mile in 3 minutes, the second in 1.5 minutes, the third in 0.75 minutes, the fourth in 0.375 minutes, the fifth in 0.1875 minutes, the sixth in 0.09375 minutes, and the seventh in 0.046875 minutes. Thus, Wivere needed 5.953125 minutes to run 7 miles. Add to this time the 6 seconds (0.1 minutes) he hesitated and Wivere reached the sea in 6.053125 minutes, or 6 minutes and about 3.4 seconds.

We can only hope that Pryor wasn't harmed by Wivere's fiery breath!

Winning at Marbles: Gopar had 12 and Dravid had 24.

Say g = the number of marbles Gopar started with and d = the number of marbles Dravid started with. So d = 2g, both at the end of the competition and before it began. After the first game, Dravid gave half of his marbles to Gopar. At that point, the number of marbles Dravid had was (½)d and the number of marbles Gopar had was g + (½)d. Following the second game, Dravid added to his (½)d marbles three-fourths of the g + (½)d marbles Gopar had after the first game. As a result Dravid had 30 marbles. Thus, (½)d + ¾[g + (½)d] = 30.

Since d = 2g, (½)d = g. Substituting g for (½)d in the above equation gives the new equation g + ¾(g + g) = 30. From this, g = 12 and d = 24.

Wizard Rankings: The latest rankings are: 1. Chameleoner, 2. Alchemerion, 3. Elvira, 4. Fortuna, 5. Bogara, 6. Deviner.

From statement (2) either Fortuna's ranking changed from 6 to 2 and Deviner's from 4 to 3, or Fortuna changed from 6 to 4 and Deviner from 4 to 6. If the former were the actual case, then Fortuna's change would have been a four-step change. But Bogara's ranking could not have changed by more than four steps, so a four-step change in Fortuna's ranking would lead to a contradiction of statement (1). So Fortuna's ranking changed from 6 to 4 and Deviner's from 4 to 6.

From statement (1), Bogara's ranking changed from 2 to 5, a three-step change. Since Elvira's change in ranking was smaller than Bogara's, her ranking changed to 3. All the rankings changed, so Alchemerion's new ranking was 2 and Chameleoner, by elimination, went to the top spot, ranking 1.

Wooden Swords: Jarius, 3; Kylar, 2. Let j = the number of matches Jarius had won and k = the number Kylar had won. Then (1) $j = k + 1$ and (2) $j + 1 = 2k$.

Substituting $k + 1$ for j in equation (2) gives $k + 2 = 2k$, from which $k = 2$. So $j = 3$.

Index

Page key: **puzzle**, *hint*, answer.

About the Author

Born near Richmond, Virginia, Margaret Edmiston received her A.B. from Upsala College, studied graduate psychology at Columbia University, and later received her M.A. in mathematics from West Georgia College. Following a long career in consumer marketing research, she now pursues writing, puzzle-making, and teaching mathematics. A longtime contributor to puzzle magazines under the names Margaret Edmiston, Margaret Shoop, and Sarah Brighton, among others, she has a continuing interest in writing for children and finds special joy and fulfillment in teaching mathematics as a member of the adjunct faculty of Virginia Commonwealth University. Ms. Edmiston is the author of *Merlin Book of Logic Puzzles* (1991), also published by Sterling.

Ms. Edmiston lives in Chesterfield County, near Richmond.